Everyday Mathematics®

The University of Chicago School Mathematics Project

MINUTE MATH+®

www.everydaymath.com

Send all inquiries to:
McGraw-Hill Education
8787 Orion Place
Columbus, OH 43240

ISBN: 978-0-02-138323-8
MHID: 0-02-138323-5

Printed in Mexico.

1 2 3 4 5 6 7 8 9 DRY 20 19 18 17 16 15

Authors
Max Bell
Jean Bell
Sheila Sconiers

Contributors
Mary Ellen Dairyko
Gina Garza-Kling
Rachel Malpass McCall
Cheryl G. Moran
Amanda Louise Ruch
Mary Fullmer
Rosalie Fruchter
Curtis Lieneck

Contents

Introduction v

Basic Routines 1

Minute Math+ Topics 23

   Operations and Algebraic Thinking 25

   Number and Operations in Base Ten 67

   Number and Operations — Fractions 95

   Measurement and Data 105

   Geometry 127

List of Activities by Page 135

Key to Sources 143

Bibliography 147

Introduction

Minute Math®+ is a collection of short mathematics activities that require little preparation. Most of the activities do not require pencil and paper, displays, measuring tools, or manipulative materials. Hence, these activities can be done anywhere. Most of them are brief enough to do in a minute.

The *Minute Math*+ activities are part of the University of Chicago School Mathematics Project's *Everyday Mathematics*® programs for Grades 1–3. They can be used with large or small groups of children at any time of the day. You might, for instance, use them during regular class time, during transitions while waiting for the group to assemble or move from one activity to another, while waiting in lines, at dismissal time, and so on. Teachers are often surprised at how much good mathematics can be learned during these few minutes.

The Role of *Minute Math*+ in the Classroom

Minute Math+ activities play at least five major roles in Grades 1–3.

1. They provide reinforcement and review of the mathematics content.
2. They provide practice with mental arithmetic and logical thinking.
3. They give children additional opportunities to think and talk about mathematics and to try out new ideas on themselves, their teachers, and their classmates.
4. They help promote the process of problem solving, which, in the long run, is more important than getting quick answers. Children become more willing to risk sharing their thoughts and their solution strategies.
5. They increase the time children spend in learning and reviewing mathematics without increasing the time spent in mathematics lessons. Since they can serve as fillers and transitions at any time during the day, they often put to good use time that would otherwise be wasted.

The Parts of *Minute Math+*

Minute Math+ is divided into six parts. You can easily identify the different parts by the tabs on the page edges.

Part 1: Basic Routines

This part of the book presents a cross-section of sample activities drawn from the parts that follow. It is a good place to begin if you have not previously used *Minute Math+* activities. As you gain experience, you can move to selecting your own mix of activities from Parts 2–6.

Parts 2–6: Minute Math+ *Topics*

Five major mathematics topics are presented in Parts 2–6:

Operations and Algebraic Thinking

Number and Operations in Base Ten

Number and Operations — Fractions

Measurement and Data

Geometry

The Organization of the Activities

Each page of *Minute Math+* begins with a basic activity followed by options for adapting that activity to various age and ability levels. The colored dot patterns in front of these options indicate a progression from easy to more difficult variations. The dots are not intended to indicate grade levels. For example:

Say: *The number is **5. Double it.*** (10)

●○○○○ Choose a 1-digit number. Tell children to double it.

●●○○○ Choose a 1-digit number. Tell children to triple it.

●●●○○ Choose a 1-digit number. Tell children to quadruple it.

●●●●○ Choose a 2-digit number. Tell children to triple or quadruple it.

The phrases and numbers in boldface type are offered as examples to help you get started. As indicated by the options, you should make substitutions depending on the age, experience, and abilities of your children.

In the example on the previous page, the activity could be stated as *The number is 5. Double it.* or *The number is 8. Triple it.* or *The number is 25. Triple it.* You can repeat the same activities many times; simply change the numbers and words as children's skills develop.

You will probably find that there are children who benefit from success at a basic level and others who appreciate being challenged at a higher level. Most of the activities involve group responses. Hence, at more challenging levels, those who are not able to respond can benefit from the responses of other children.

Many of the *Minute Math+* activities use real data about people, animals, nature, and other topics. These numbers are usually interesting and often surprising. A key to sources is provided at the end of the book.

The most important things for you to remember as you use *Minute Math+* are that the activities should meet the needs of your children and they should also be catalysts for your own ideas. Ask the children for their activity ideas. Try new things and larger numbers—the children's responses may surprise you.

Basic Routines

Notes

Numbers Before and After

Ask: *What number comes after (follows)* **8** *when you count?*

What number comes before (precedes) **4** *when you count?*

What numbers come before and after **6**?

⦿○○○○ Use 1-digit numbers.

⦿⦿○○○ Substitute 2-digit numbers, including those that will change the decade. Example: *What number comes after 29 when you count?* (30)

⦿⦿⦿○○ Ask the child what number comes 2 or 3 (or another number) *before* or *after* a given number. Example: *What number comes 2 before 9 when you count?* (7)

⦿⦿⦿⦿○ Substitute 3-digit or 4-digit numbers and ask what number comes 2 or 3 (or another number) *before* or *after* the given number. Examples: *What number comes 3 after 391 when you count?* (394) *What number comes 3 before 391 when you count?* (388)

CCSS **1.NBT.1, 2.NBT.2, 3.NBT.2**

Numbers Between

Say: *Tell me any numbers between **25** and **35**.*

⬤◯◯◯◯ Choose whole numbers.

⬤⬤◯◯◯ Choose negative numbers. Example: *Tell me any numbers between 0 and −4.*

⬤⬤⬤◯◯ Choose fractions or decimals. Example: *Tell me any numbers (fractions) between $\frac{1}{4}$ and $\frac{3}{4}$.* (An infinite number of answers are possible.)

Counts and Skip Counts

Say: *I will begin counting. If I point to you, continue the number sequence until I stop you and point to someone else.* **15, 20, 25, 30, . . .**

- ●○○○○ Begin with numbers 1–50. Count by 2s, 5s, or 10s.

- ●●○○○ Begin with numbers 1–100. Count forward or backward by 1s, 2s, 5s, or 10s.

- ●●●○○ Begin with numbers 100–900. Count forward or backward by 1s, 2s, 5s, 10s, 25s, or 50s.

- ●●●●○ Begin with negative numbers or with numbers greater than 1,000. Count forward or backward by 1s, 2s, 5s, 10s, 25s, 50s, or 100s.

Basic Routines

Ordinal Numbers

Ask: *Who is **second** in line? Who is **10th** in line?*

○○○○○ Ask: *Who is first in line? Who is third from the end? Who is third from the beginning? What number is that person from the end? Who is eighth in line?*

●○○○○ Identify children by their place in line, and ask them to perform some sort of action. Example: Have the third child in line and the seventh child in line change places. Ask: *Is the fourth child in line still the same?*

Variation 1 Perform several actions such as jumping, clapping hands, and snapping fingers in sequence. Ask: *What did I do **first? Next? Last?*** (or ***first, second, third***) Ask a volunteer to perform another sequence of actions.

Variation 2 *Which letter is **12th** in the alphabet?* (The letter L) *The letter **T** is what letter?* (The 20th letter)

Count by 10s and 100s

Say: *Count with me: 13, 23, 33, 43, . . .*

○○○○○ Count forward and backward by 10s. Begin with any 2-digit number.

●●○○○ Count forward and backward by 10s or 100s. Begin with any 3-digit number.

●●●○○ Count forward and backward by 10s, 100s, or 1,000s. Begin with any 4-digit number.

●●●●○ Start with any number. Count forward and backward by any power of 10.

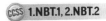

What Do I Do?

Say: *I have **4**, but I want **7**. What do I have to do?*

Remember to provide a context for the numbers.

- ●○○○○ Create problems using small whole numbers. Example: *I have 6, but I only want 3. What do I have to do?* (Subtract, or take away, 3.)

- ●●○○○ Create problems using whole numbers 1–100 divisible by 5 or 10. Example: *I have 10, but I want 30. What do I have to do?* (Add 20.)

- ●●●○○ Create problems using simple fractions. Example: *I have $\frac{1}{4}$, but I want 1 whole. What do I have to do?* $\left(\text{Add } \frac{3}{4}, \text{ or get another } \frac{3}{4}.\right)$

- ●●●●○ Create problems using whole numbers 1–100. Example: *I have 42, but I need 64. What do I have to do?* (Add 22.)

Using Combinations of 10

Ask: *What number must you add to **6** to get **10**?*

⦿○○○○ Say a 1-digit number and ask what must be added to get 10.
Example: *What must you add to 3 to get 10?* (7)

⦿⦿○○○ Say a 1- or 2-digit number and ask what must be added to get to
another decade. Examples: *What must you add to 26 to get 30?* (4)
What must you add to 26 to get 50? (24)

⦿⦿⦿○○ Say a 1- or 2-digit number and ask what must be added to get to
the next century. Examples: *What must you add to 6 to get 100?*
(94) *What must you add to 26 to get 100?* (74)

⦿⦿⦿⦿○ Say a 3-digit number and ask what must be added to get to the
next 10, 100, or 1,000. Examples: *What must you add to 362 to get
370?* (8) *What must you add to 362 to get 400?* (38) *What must you
add to 362 to get 1,000?* (638)

Arithmetic Facts

Ask: **4** + **6** makes how much? What does **7** − **3** equal?

Remember to provide a context for the numbers.

◑○○○○ Use single-digit addition and subtraction facts.
Example: *9 plus 1 makes how much?* (10)

◑◑○○○ Use 2-digit addition and subtraction facts where at least one number is divisible by 10.
Example: *What is 50 plus 26?* (76)

◑◑◑○○ Use single-digit multiplication and division facts.
Example: *What does 6 times 4 equal?* (24)

◑◑◑◑○ Use 2-digit addition and subtraction fact extensions.
Examples: *If 5 + 7 = 12, how much is 5 + 27?* (32)
How much is 5 + 87? (92) *How much is 5 + 687?* (692)

Name Collections (Equivalents)

Say: *I'm thinking of the number 10. What are other names for 10?*
$(5 + 5, 11 - 1, 20 \div 2, 1 + 9)$

●○○○○ Think of a whole number 1–10. Ask for other names for it.

●●○○○ Think of a whole number 10–100. Ask for other names for it.

●●●○○ Think of a fraction. Ask for other names for it.

●●●●○ Think of a negative number. Ask for other names for it.

Variation 1 Think of two (or three or four) numbers that, added together, make 10. Ask what they could be. $(6 + 4, 5 + 5,$ and so on$)$

Variation 2 Think of two numbers that have a difference of 10. Ask what they could be. $(12 - 2, 100 - 90,$ and so on$)$

Variation 3 Think of a name for 10 that involve products, quotients, or fractions. Ask what it could be. $\left(\frac{100}{10}, 10 \times 1, \frac{1}{2}\right.$ of 20, and so on$\left.\right)$

CCSS **1.OA.6, 1.OA.7, 2.OA.2, 2.NBT.5, 3.OA.7, 3.NF.3b, 3.NF.3c**

More Name Collections (Equivalents)

Say: *I'm thinking of 50¢. What are some other names for 50¢?*
(2 quarters, 5 dimes, 10 nickels, and so on)

○●○○○ Think of a money amount. Ask for other names for it.

●●○○○ Think of a metric measure. Ask for other names for it. Example:
What are some other names for $\frac{1}{2}$ meter? (50 cm, $\frac{1}{2}$ of 100 cm, and
so on)

●●●○○ Think of a fraction. Ask for other names for it. Example: *What are
some other names for $\frac{1}{2}$?* $\left(\frac{2}{4}, \frac{3}{6}, \frac{10}{20}, \text{and so on}\right)$

●●●●○ Think of a large decade or century number. Ask for other names
for it. Example: *What are some other names for 4,000?*
(40 hundreds, 400 tens, and so on)

●●●●● Think of decimal numbers. Ask for other names for it. Example:
What are some other names for 0.01? $\left(\frac{1}{100}, \frac{10}{1,000}, \text{and so on}\right)$

Multistep Problems

Say: *Listen carefully to each step of this problem. Raise your hand when you have the answer. **9 take away 3, plus 4, add 2, minus 1 makes how much?*** (11)

Remember to provide a context for the numbers.

- ●○○○○ Use addition and subtraction of small whole numbers. Example: *Add 5 to 9, subtract 2, add 12, add 6, subtract 15 equals what?* (15)

- ●●○○○ Use addition and subtraction and doubling of 2-digit whole numbers. Numbers divisible by 5 or 10 are easiest. Example: *Subtract 10 from 25, add 5, double it, minus 10 equals what?* (30)

- ●●●○○ Use addition, subtraction, multiplication, doubling, and halving of small whole numbers. Example: *Add 12 and 10, halve it, take away 4, times 2 equals what?* (14)

Variation If calculators are available, young children might practice some problems of this sort on the calculators.

CCSS **1.OA.6, 1.NBT.4, 2.OA.2, 2.NBT.5, 3.OA.7, 3.NBT.2**

How Many 10s, 100s, 1,000s?

Ask: *How many **10s** are in **100**?* (10)

⦿○○○○ Ask about tens in hundreds. Example: *How many 10s in 200?* (20) *How many 10s in 800?* (80)

⦿⦿○○○ Ask about hundreds in thousands. Example: *How many 100s in 1,000?* (10) *In 3,000?* (30)

⦿⦿⦿○○ Ask about hundreds in ten-thousands. Example: *How many 100s in 10,000?* (100) *In 25,000?* (250)

⦿⦿⦿⦿○ Ask: *How many 1,000s are in 1 million?* (1,000)

"What's My Rule?"

Say: *The rule is **add 5**. If the input is **6**, what is the output?* (11)

●○○○○ Give a rule that adds or subtracts a number 1–10. Give the input and ask for the output.

●●○○○ Give a rule that adds or subtracts a 2-digit number. Give the input and ask for the output.

●●●○○ Give a rule that multiplies or divides by a number 1–10. Give the input and ask for the output.

●●●●○ Give a rule that takes a fraction or a percent of another number. Example: *The rule is take $\frac{1}{2}$. If the input is 14, what is the output?* (7)

Variation 1 Give the input and the output. Ask children to supply the rule. Example: *If the input is 8 and the output is 11, what is the rule?* (+3, or add 3.)

Variation 2 Give the rule and the output. Ask children to supply the input. Example: *If the rule is add 7 and the output is 12, what is the input?* (5)

Number Stories   Number Stories

Say: *If I choose you, select two numbers and ask someone else to make up an **addition number story** using those two numbers.*

- ●○○○○ The chosen child asks another child to make up an addition number story.

- ●●○○○ The chosen child asks another child to make up a subtraction number story.

- ●●●○○ The chosen child asks another child to make up a multiplication number story.

- ●●●●○ The chosen child asks another child to make up a number story with an answer that fits a certain characteristic they select. For example, the chosen child asks someone else to make up a number story for which the answer is a negative number or for which the answer is more than 15.

Shapes Around Us

Ask: *What **2-dimensional** shapes do you see **in this room?***
(Circles, triangles, squares, rectangles, and so on)

Ask: *What **3-dimensional** shapes do you see **in this room?***
(Spheres, pyramids, cylinders, cubes, prisms, cones, and so on)

Also have children look for shapes in the hall, on the playground, and in other locations. On field trips, have children watch for and "collect" (in their minds) as many of a particular shape as possible. After the trip, children discuss the shapes they found.

Geometry "I Spy"

Say: *I will choose a "spy." That spy will choose an object in the room and tell us what shape it is. For example, the spy might say, "I spy a **circle.**" We will ask the spy yes or no questions to try to determine the object.*

●○○○○ Prompt children to "spy" 2-dimensional shapes: circles, triangles, squares, rectangles, and so on.

●●○○○ Prompt children to "spy" 3-dimensional shapes: spheres, pyramids, cylinders, cubes, prisms, or cones.

How Many Cents?

Ask: *How many cents do I have if I have **3 quarters?***

○●○○○○ Use 1–10 of the same coin. Example: *How many cents do I have if I have 6 nickels?* (Count by 5s to reach 30¢.)

●●○○○ Combine 2 coins or use more than 10 of the same coin. Example: *How many cents do I have if I have 5 dimes and 2 nickels?* (Count by 10s to reach 50¢, and then count on 10 more by 5s to reach 60¢.) *How many cents do I have if I have 20 nickels?* (Count by 5s to reach 100¢, or $1.)

●●●○○ Combine 3 coins. Example: *How many cents do I have if I have 2 dimes, 3 nickels, and 26 pennies?* (61¢) Discuss strategies.

●●●●○ Use bills. Examples: *How many cents do I have if I have a ten-dollar bill?* (1,000¢) *How many cents do I have if I have a five-dollar bill?* (500¢)

Place Value

Say: *Tell me a **2-digit** number with **8** in the **tens place.** Tell me another number.*

●○○○○ Work with 2-digit numbers. Request specific numbers in the tens place or ones place.

●●○○○ Work with 3-digit numbers. Request specific numbers in the tens place or hundreds place.

●●●○○ Work with 4-digit numbers. Request specific numbers in the tens, hundreds, or thousands place.

Finding 10 More and 10 Less

Ask: *What number is 10 more than **9**? What number is 10 less than **23**?*

⦿○○○○ Use 1-digit numbers with 10 more and 10 less.

⦿⦿○○○ Use 2-digit numbers with 10 more and 10 less or with 100 more and 100 less.

⦿⦿⦿○○ Use 3-digit numbers with 10, 100, or 1,000 more and less.

⦿⦿⦿⦿○ Use 4-digit (or more) numbers with 1,000 or 10,000 more and less.

⦿⦿⦿⦿⦿ Start with negative numbers or let the answer be negative. Examples: *What is 10 more than −25? (−15) What is 10 less than 8? (−2)*

Minute Math+ Topics

Operations and Algebraic Thinking 25

Number and Operations in Base Ten....................... 67

Number and Operations — Fractions..................... 95

Measurement and Data 105

Geometry 127

Notes

Missing Parts in Sums and Differences

Ask: *What do you subtract from **12** to get **4**?* (8) *From what number do you subtract **3** to get **6**?* (9) *What do you add to **5** to get **8**?* (3) (Vary the format of the questions, and ask children to discuss their strategies.)

Remember to provide a context for the numbers.

◐○○○○ Create problems using small whole numbers. Examples: *What do you subtract from 9 to get 9?* (0) *What do you add to 4 to get 7?* (3) *From what number do you subtract 4 to get 4?* (8)

◐◐○○○ Create problems using whole numbers that are divisible by 5 or 10. Example: *What number do you subtract from 50 to get 40?* (10)

◐◐◐○○ Create problems using simple fractions. Example: *What do you add to $\frac{1}{4}$ to get 1 whole?* $\left(\frac{3}{4}\right)$

◐◐◐◐○ Create problems using whole numbers 10–100.

◐◐◐◐◐ Create problems using money or time.

Sleep Needs `Number Stories`

A growing child needs about 10 hours of sleep each night.

●○○○○ If Jorge has been asleep for about 4 hours, how many more hours should he sleep? (6 hr) If Kim has been asleep for about 7 hours, how many more hours should she sleep? (3 hr)

●●○○○ If Alexis goes to bed at 9:00 P.M., at about what time should she get up in the morning? (7:00 A.M.)

●●●○○ If Rachel wants to be up by about 6:00 A.M. to go fishing, at about what time should she go to bed? (8:00 P.M.)

●●●●○ About how many minutes of sleep does a growing child need each night? (600 min)

Thunder   Number Stories

It takes about 5 seconds for the sound of thunder to travel 1 mile.

●○○○○ About how long would it take the sound of thunder to travel
2 miles? (10 sec) About how long would it take the sound of
thunder to travel 5 miles? (25 sec) 10 miles? (50 sec)

●●○○○ About how far can the sound of thunder travel in 1 minute? (12 mi)
About how far can the sound of thunder travel in $\frac{1}{2}$ minute? (6 mi)

Rug Measures Number Stories

The rug in Joshua's room is about **8** feet wide. It is about **2** feet longer than it is wide.

⦿○○○○ About how long is the rug? (10 ft)

⦿⦿○○○ About how many inches wide is the rug? (96 in.) About how many inches long is the rug? (120 in.)

⦿⦿⦿○○ What is the area of the rug? (80 sq ft)

Old Milk   Number Stories

Milk spoils about 1 week after the expiration date on the package.

● ○ ○ ○ ○   About how many days after the expiration date does milk spoil? (7 days) If you buy 2 containers of milk, about how long after the expiration date will they spoil? (Still about 1 week)

● ● ○ ○ ○   If the expiration date on Mr. Mayer's milk is 2 days ago, in about how many more days will it spoil? (5 days) If the expiration date is 6 days ago, in about how many days will the milk spoil? (1 day)

● ● ● ○ ○   If today is **Wednesday** and Mr. Mayer's milk will spoil in about 3 days, on what day will it begin to spoil? (Saturday)

● ● ● ● ○   If the expiration date on the milk is January 26, on what day will the milk probably spoil? (February 2)

Operations and Algebraic Thinking

Dreams Number Stories

People dream an average of 5 times per night.

●○○○○ At that average, about how many dreams might you have tonight?
(5 dreams) About how many dreams would you have had last night
and the night before? (10 dreams)

●●○○○ About how many dreams might you have had since **Tuesday**?
About how many dreams do you have in 1 week? (35 dreams)

●●●○○ About how many dreams might our whole class have on some
Tuesday night? (Number of children × 5)

●●●●○ About how many dreams might you have in 1 month? (150 dreams)
About how many dreams might you have in 1 year? (1,800 dreams)

Classroom Counts   Number Stories

How many children are in the classroom today?

●○○○○ How many children are absent? If each child receives a carton of milk for lunch, how many cartons of milk do we need for our class?

●●○○○ If we invite the class next door over for lunch, and they have the same number of children that we have, how many cartons of milk do we need?

●●●○○ If there are _____ classrooms in our school, about how many children are there in our school?

●●●●○ If there are _____ classes for each grade and _____ grades in our school, how many classrooms are in our school? About how many children are in our school? If each child drinks milk at lunch, about how many cartons of milk should be ordered each day?

Hot Dog Buns   Number Stories

Hot dogs often come in packages of 10. Buns usually come in packages of 8.

- ●○○○○ For a class picnic, would we need to buy more packages of hot dogs or packages of buns? (More packages of buns) Why? (There are fewer buns per package.)

- ●●○○○ If we want to buy enough hot dogs and buns for 20 people, how many packages of each do we need to buy? (2 packages hot dogs, 3 packages buns) For 35 people? (4 packages hot dogs, 5 packages buns) How many of each would be left over? (5 hot dogs, 5 buns)

- ●●●○○ What is the least number of packages of hot dogs and buns you would have to buy to have the same number of each? (4 packages hot dogs, 5 packages buns)

- ●●●●○ For 5 packages of hot dogs, how many packages of buns are needed? (7 packages) How many of each would we have? (50 hot dogs, 56 buns) For 8 packages of hot dogs, how many packages of buns do we need? (10 packages) How many of each would we have? (80 hot dogs, 80 buns)

Baby Penguin Meals Number Stories

Baby penguins eat almost all the time. Penguin parents feed a baby about 2 pounds of food every hour.

● ○ ○ ○ ○ About how much does a baby penguin eat in 2 hours? (4 lb) In 5 hours? (10 lb)

● ● ○ ○ ○ A baby penguin has eaten about 10 pounds of food today. About how many hours have passed? (5 hr)

● ● ● ○ ○ About how much does a baby penguin eat in 1 day (24 hours)? (48 lb) In 2 days? (96 lb)

Operations and Algebraic Thinking

Fact Families

Say: *Here are 3 numbers of a fact family: **2, 8, and 10.** What are all the* ***addition*** *and* ***subtraction*** *facts of the family?*
$(2 + 8 = 10, 8 + 2 = 10, 10 - 8 = 2,$ and $10 - 2 = 8)$

- ●○○○○ Give children 1-digit numbers from an addition/subtraction fact family.

- ●●○○○ Give children 1-digit numbers from a multiplication/division fact family. Example: Give children 2, 8, and 16.
$(2 \times 8 = 16, 8 \times 2 = 16, 16 \div 8 = 2,$ and $16 \div 2 = 8)$

- ●●●○○ Give children numbers from a fraction fact family. Example: Give children $\frac{1}{4}, \frac{3}{4},$ and 1. $\left(\frac{1}{4} + \frac{3}{4} = 1, \frac{3}{4} + \frac{1}{4} = 1, 1 - \frac{1}{4} = \frac{3}{4},$ and $1 - \frac{3}{4} = \frac{1}{4}\right)$

Making Omelets **Number Stories**

Most omelets require 3 eggs. For 3-egg omelets:

⬤○○○○ How many eggs would you need to make 2 omelets? (6 eggs) To make 3 omelets? (9 eggs)

⬤⬤○○○ How many omelets can you make from 1 dozen eggs? (4 omelets) From 3 dozen eggs? (12 omelets)

⬤⬤⬤○○ How many dozen eggs would you use in 1 omelet? $\left(\frac{1}{4}\text{ dozen}\right)$ In 2 omelets? $\left(\frac{1}{2}\text{ dozen}\right)$

Operations and Algebraic Thinking

Bags of Apples Number Stories

A small bag of apples contains 5 apples.

●○○○○ If Marcia eats 1 apple from the bag and gives another apple to David, how many apples will be left in the bag? (3 apples) If she gives Karen one of the remaining apples, how many will be left? (2 apples)

●●○○○ How many apples are in 3 bags? (15 apples) In 5 bags? (25 apples)

●●●○○ If Greg has 7 apples, how many bags does he have? (At least 2 bags) If Bob has 16 apples, how many bags does he have? (At least 4 bags)

Toilet Flushes `Number Stories`

Suppose each toilet flush uses about 7 gallons of water.

○○○○○ About how many gallons of water do 2 toilet flushes use? (14 gal)

○○○○○ About how many gallons of water do 3 flushes use? (21 gal)
4 flushes? (28 gal)

○○○○○ There are 4 quarts in 1 gallon. About how many quarts of water
does each flush use? (28 qt) About how many quarts of water do
2 flushes use? (56 qt) There are 8 pints in 1 gallon. About how
many pints of water does each flush use? (56 pt)

○○○○○ How many times would we need to flush a toilet to use about
112 pints of water? (2 times)

○○○○○ If there are 5 people in my family and each person flushes the
toilet 3 times a day, about how many gallons of water are used
each day? (105 gal)

Operations and Algebraic Thinking

A Snail's Pace   Number Stories

The slowest snails in the world move at a speed of about 23 inches per hour.

●○○○○   Do these snails travel more or less than 1 foot in an hour? (More)
         Do they travel more or less than 2 feet in an hour? (A little less)

●●○○○   About how many inches do these snails travel in 2 hours? (46 in.)
         In half an hour? (11 or 12 in.)

●●●○○   About how long would it take this kind of snail to get across a
         table that is 4 feet wide? (About 2 hr) To cross a room that is
         12 feet wide? (About 6 hr)

Making Apple Juice Number Stories

As a rule of thumb, you need 3 apples to make 1 glass of apple juice.

●○○○○ About how many apples would you need to make 2 glasses of juice? (6 apples) To make 3 glasses of juice? (9 apples)

●●○○○ About how many apples would we need to make a glass of juice for each child in this classroom? (Number of children × 3)

●●●○○ If we have 1 dozen apples, about how many glasses of juice could we make? (4 glasses) If we have 2 dozen apples, about how many glasses of juice could we make? (8 glasses)

Operations and Algebraic Thinking

Baby Weights Number Stories

The average newborn human baby doubles its weight in 6 months.

○○○○○ About how much will a baby weigh in 6 months if it weighed about 7 pounds at birth? (14 lb) If it weighed about 8 pounds? (16 lb)

●●○○○ If a 6-month-old baby weighs about 12 pounds, about how much did it weigh at birth? (6 lb) If a 6-month-old baby weighs about 14 pounds, about how much did it weigh at birth? (7 lb) If it weighs about 16 pounds? (8 lb)

●●●○○ If a newborn weighs about 7 pounds, about how much will it weigh when it is 2 years old? (Insufficient data to answer the question; you don't know how much weight an infant will gain after the first 6 months.) If a 6-pound newborn did keep doubling its weight every 6 months, about how much would it weigh when it is 3 years old? (Almost 400 lb; that is more than twice the average adult male.)

Cooking-Oil Consumption [Number Stories]

Suppose that a cafeteria uses about 6 quarts of cooking oil every day.

○●○○○○ About how many quarts of cooking oil does the cafeteria use in 2 days? (12 qt) In 3 days? (18 qt) In 10 days? (60 qt)

●●○○○ About how many quarts of cooking oil does the cafeteria use each week? (About 42 qt) About how many quarts of oil does the cafeteria use in a month? (About 180 qt or 186 qt)

●●●○○ There are 2 pints in 1 quart. About how many pints of cooking oil does the cafeteria use each week? (84 pt) There are 4 quarts in 1 gallon. About how many gallons of cooking oil does the cafeteria use each week? $\left(10\frac{1}{2}\text{ gal}\right)$

●●●●○ Would 15 gallons of cooking oil be enough for 2 weeks? (No.) Would 25 gallons of cooking oil be enough for 2 weeks? (Yes.)

Operations and Algebraic Thinking

Planting Flower Bulbs `Number Stories`

As a rule of thumb, plant a flower bulb three times as deep as its length.

●○○○○ About how many inches deep should you plant a 1-inch bulb? (3 in.) Show me, with your hands, about how deep this would be. About how many inches deep should you plant a 2-inch bulb? (6 in.)

●●○○○ If I followed this rule and planted a bulb 6 centimeters deep, how long is the bulb? (2 cm) If I planted the bulb 9 centimeters deep? (3 cm)

●●●○○ About how many inches deep should you plant a bulb that is $1\frac{1}{2}$ inches long? $\left(4\frac{1}{2} \text{ in.}\right)$ That is $\frac{1}{2}$ inch long? $\left(1\frac{1}{2} \text{ in.}\right)$

Digit Arithmetic

Say: *Think of a 2-digit number in which **the sum of the digits is 11**.* (65, 56, 74, 47, 83, 38, 92, 29)

●○○○○ The sum of the two digits equals a number 1–18. Also ask for 3- or 4-digit numbers the sum of whose digits is a number of your choice. Example: *What is a 3-digit number whose digits have a sum of 3?* (111, 210, 201, 102, 120, 300)

●●○○○ The difference of the two digits equals a number 0–9. Example: *Think of a 2-digit number in which the difference between the two digits equals 5.* (16, 61, 27, 72, 38, 83, 49, 94, 50)

●●●○○ The product of the two digits equals a number 0–81. Example: *Think of a 2-digit number in which the product of the two digits equals 18.* (29, 92, 36, 63)

Elephant Sleep Number Stories

Most elephants need only about 2 hours of sleep each day.

●○○○○ About how many hours of sleep do you need each day? About how many more hours of sleep do you need than most elephants?

●●○○○ About how many hours per week does an elephant sleep? (About 14 hr) About how many hours per month does an elephant sleep? (About 60 hr)

●●●○○ About how many minutes per day does an elephant sleep? (About 120 min) About how many minutes does an elephant sleep in 2 days? (About 240 min)

●●●●○ About how many hours per year does an elephant sleep? (About 730 hr) About how many hours per year do you sleep?

Hibernation Number Stories

The burrow ground squirrel, which lives in Alaska, hibernates for about 9 months of the year.

- ●○○○○ For about how many months of the year is the burrow ground squirrel active (not hibernating)? (3 months)

- ●●○○○ For about how many months does this squirrel hibernate in 2 years? (18 months) In 3 years? (27 months) During your lifetime?

- ●●●○○ For about how many days of the year does the burrow ground squirrel hibernate? (About 270 to 280 days) For about how many days of the year is the burrow ground squirrel not hibernating? (About 90 to 93 days)

- ●●●●○ For about what fraction, or portion, of the year does the burrow ground squirrel hibernate? $\left(\frac{9}{12} \text{ or } \frac{3}{4} \text{ of the year}\right)$

Variation Pose questions about the Siberian chipmunk, which hibernates 7 to 8 months of the year.

Operations and Algebraic Thinking

Making Muffins Number Stories

A muffin recipe calls for 2 cups of flour and 1 cup of sugar for each batch.

●○○○○ If you wanted to make twice as many muffins, how many cups of sugar should you use? (2 cups) How many cups of flour? (4 cups)

●●○○○ If you wanted to make half as many muffins, how many cups of sugar would you use? $\left(\frac{1}{2} \text{ cup}\right)$ How many cups of flour? (1 cup)

●●●○○ Using a half-cup measuring cup, how many half cups of flour do you need for one batch of muffins? (4 half cups) How many half cups of sugar? (2 half cups) Using a one-third cup measuring cup, how many one-third cups of flour would you need? (6 one-third cups) How many one-third cups of sugar? (3 one-third cups)

●●●●○ If you want to make twice as many muffins as the recipe makes, and use a half-cup measuring cup, how many half cups of flour would you need to use? (8 half cups) How many half cups of sugar? (4 half cups)

Types of Bears Number Stories

Scientists have identified eight different types of bears in the world. Three of these different types can be found in North America.

●○○○○ How many types of bears in the world cannot be found in North America? (5 types)

●●○○○ What fraction, or portion, of the types of bears in the world can be found in North America? $\left(\frac{3}{8}\right)$ What fraction, or portion, of the types of bears in the world cannot be found in North America? $\left(\frac{5}{8}\right)$

Pumping Blood [Number Stories]

As a rule of thumb, your heart pumps about 5 tablespoons of blood into your arteries with each beat.

●○○○○ About how many tablespoons of blood are pumped into your arteries in 2 beats? (10 tbs) In 3 beats? (15 tbs) In 10 beats? (50 tbs)

●●○○○ If your heart beats about 60 times per minute, about how many tablespoons of blood does your heart pump into your arteries in 1 minute? (300 tbs) In 2 minutes? (600 tbs)

●●●○○ If you were chasing a friend, would your heart pump more or less blood into your arteries in one minute than if you were reading a book? Explain. (More blood; when you are active, your heart beats more rapidly than when you are relaxed. Since this means your heart beats faster when you run, more tablespoons of blood would be pushed into your arteries.)

Basketball Player Heights* `Number Stories`

In 1989, the tallest basketball player in the National Basketball Association (NBA) was Manute Bol. He measured 7 feet 6 inches in height. The shortest player in the NBA was Tyrone Bogues, who measured a little more than 5 feet 3 inches in height.

○○○○○ Was the difference in height between these players more or less than 1 foot? (More) Was the difference in height more or less than 2 feet? (More) More or less than 3 feet? (Less)

●●○○○ About how many feet and inches over 6 feet tall is Manute Bol? (1 ft 6 in.) About how many inches under 6 feet tall is Tyrone Bogues? (9 in.)

●●●○○ About how much shorter is Tyrone Bogues than Manute Bol? (2 ft 3 in.)

●●●●○ About how many inches tall is Manute Bol? (90 in.) Tyrone Bogues? (63 in.)

*You may want to display the heights of the players.

Walking Rates Number Stories

The average city dweller walks about 6 feet per second while the average country dweller walks about 3 feet per second.

⦿○○○○ Who will walk farther in 1 minute: the average city dweller or the average country dweller? (City dweller) How many more feet per second does the average city dweller walk than the average country dweller? (3 ft per sec)

⦿⦿○○○ About how many feet does the average city dweller walk in 2 seconds? (12 ft) In 5 seconds? (30 ft) About how many feet does the average country dweller walk in 2 seconds? (6 ft) In 5 seconds? (15 ft)

⦿⦿⦿○○ In 1 minute, does the average city dweller walk more or less than 200 feet? (More) The average country dweller? (Less) In 2 minutes, about how far does the average country dweller walk? (360 ft)

⦿⦿⦿⦿○ In 1 minute, about how many yards does the average city dweller walk? (120 yd) The average country dweller? (60 yd)

Paper in Trash Heaps Number Stories

Some people claim that the average 10-ton pile of municipal trash contains about 3 tons of paper.

●○○○○ Is more or less than half of the pile paper? (Less) If people begin to recycle more paper, what will happen to the amount of paper in municipal trash? (It will decrease.)

●●○○○ About how many tons of paper are in a 20-ton pile of trash? (6 tons) In a 30-ton pile of trash? (9 tons) In a 40-ton pile? (12 tons)

●●●○○ If there are 20 tons of trash one day and 50 tons of trash a few days later, about how many tons of paper have been added? (9 tons) Since there are 2,000 pounds in 1 ton, about how many pounds of paper have been added? (18,000 lb)

●●●●○ If there are 19 tons of paper in the trash heap, about how much does the whole trash heap weigh? (About 60 tons) Since there are 2,000 pounds in 1 ton, about how much does the trash heap weigh in pounds? (120,000 lb)

Operations and Algebraic Thinking

Secret Numbers

Say: *Each of you think of a secret number 1–10. When I call on you, **add 22** to your secret number, tell us only the answer, and call on someone else to guess your secret number. For instance, if **Debbie's** secret number is 4, she will say 26, and then call on someone to guess her secret number.*

●○○○○ Choose a 1- or 2-digit number to add to or subtract from the secret number. Example: Add 23 to the secret number.

●●○○○ Choose a 1-digit number to multiply by the secret number. Example: Multiply 8 by the secret number.

●●●○○ Choose two numbers; have the child add the first one to the secret number and then subtract the second one. Example: Add 31 to the secret number, and then subtract 10.

Variation Provide the whole class with an operation, and then have children take turns with a partner thinking of and calculating the secret number.

Penguins vs. Human Swimmers Number Stories

Penguins can swim about 15 miles in 1 hour. A fast human can swim about 3 miles in 1 hour.

●○○○○ About how many more miles than a human can a penguin swim in 1 hour? (12 mi)

●●○○○ About how far could a penguin swim in 2 hours? (30 mi) A human? (6 mi; however, it is unlikely that a human could actually swim fast for 2 hours.) About how far could a penguin swim in $\frac{1}{2}$ hour? $\left(7\text{–}8 \text{ mi, or about } 7\frac{1}{2} \text{ mi}\right)$ A human? $\left(1\text{–}2 \text{ mi, or } 1\frac{1}{2} \text{ mi}\right)$

●●●○○ About how many times faster than a human can a penguin swim? (About 5 times faster)

●●●●○ About how long would it take a penguin to swim 3 miles? $\left(\frac{1}{5} \text{ hr, or about 12 min}\right)$ If a human could continue swimming at a very fast pace, how long would it take the swimmer to swim 15 miles? (At least 5 hr)

CCSS 1.OA.1, 1.NBT.4, 2.OA.1, 2.MD.5, 3.OA.3, 3.NF.1

Operations and Algebraic Thinking

Animal Litters   Number Stories

Hamsters can have litters about once every 4 months. Each litter usually consists of 6–12 baby hamsters.

●○○○○ About how many litters can hamsters have in 8 months? (2 litters) In 13 months? (3 litters)

●●○○○ About how many times per year can a hamster have a litter? (3 times) About how many times in 2 years can a hamster have a litter? (6 times)

●●●○○ About how many babies can a hamster have in 1 year? (18–36 babies)

Variation   Pose questions about the birth rates of other animals: red squirrels have 1–3 litters per year with 3–7 young per litter; wood mice have 3–4 litters per year with 4–6 young per litter; foxes average 1 litter per year with 3–8 young per litter.

Animal Weights* `Number Stories`

The average rat weighs about 1 pound. The average chicken weighs about 7 pounds. The average cat weighs about 14 pounds.

- ⬤○○○○ About how much heavier is a chicken than a rat? (6 lb) About how much lighter is a chicken than a cat? (7 lb) A rat than a cat? (13 lb)

- ⬤⬤○○○ About how much would 2 chickens weigh together? (14 lb) About how much would 1 cat and 1 rat weigh together? (15 lb)

- ⬤⬤⬤○○ If we put 1 chicken on one side of a balance, how many rats on the other side would make it balance? (7 rats) If we put 2 chickens on one side of a balance, how many rats on the other side would make it balance? (14 rats) If we put 1 cat on one side of a balance, how many chickens on the other side would make it balance? (2 chickens)

- ⬤⬤⬤⬤○ Which would weigh more, 15 rats or 2 chickens? (15 rats) Which would weigh less, 2 cats or 2 chickens and 5 rats? (2 chickens and 5 rats)

*You may want to display the weights of the animals.

CCSS 1.OA.1, 1.OA.6, 2.OA.1, 2.NBT.5, 3.OA.7, 3.NBT.2

Growing Eyelashes   Number Stories

On the average, a person grows a new eyelash every 3 months.

●○○○○ About how many months would it take to grow 2 new eyelashes?
(6 months) To grow 3 new eyelashes? (9 months)

●●○○○ On the average, about how many new eyelashes does a person
grow in one year? (4 new eyelashes) In 2 years? (8 new eyelashes)

●●●○○ On the average, about how many weeks does it take to grow a
new eyelash? (About 12 weeks) To grow 4 new eyelashes? (About
48 weeks)

●●●●○ On the average, about how many days does it take to grow 10 new
eyelashes? (About 900 days)

Milk Consumption `Number Stories`

Imagine that a nursery school uses about 5 quarts of milk every day.

⬤○○○○ About how many quarts of milk do they use in 2 days? (10 qt)
In 4 days? (20 qt)

⬤⬤○○○ About how many quarts of milk do they use in 5 days? (25 qt)
In 7 days? (35 qt) In 10 days? (50 qt)

⬤⬤⬤○○ There are 2 pints in 1 quart. About how many pints of milk do they
use each day? (10 pt) In 3 days? (30 pt)

⬤⬤⬤⬤○ If each child drinks about a half pint of milk each day, how many
children are in the nursery school? (20 children)

Watching Sheep   Number Stories

A sheep farmer needs 1 sheepdog for every 100 sheep.

⦿○○○○ How many sheepdogs does the farmer need to watch 200 sheep? (2 sheepdogs) To watch 300 sheep? (3 sheepdogs)

⦿⦿○○○ How many sheepdogs does the farmer need to watch a flock of 500 sheep? (5 sheepdogs) A flock of 1,000 sheep? (10 sheepdogs)

⦿⦿⦿○○ A sheep farmer has 4 sheepdogs. About how many sheep does this farmer have? (400 sheep) If a farmer has 7 sheepdogs, about how many sheep does the farmer have? (700 sheep)

⦿⦿⦿⦿○ A farmer has 300 sheep, but only 1 sheepdog. How many more sheepdogs will the farmer need? (2 more sheepdogs) If a farmer has 600 sheep and only 2 sheepdogs, how many more sheepdogs will he need? (4 more sheepdogs)

Granola Bar Sale Number Stories

Suppose that the grocery store is selling granola bars 3 for $1.00.

◐○○○○ How many granola bars could you buy for $2.00? (6 granola bars)
How many granola bars could you buy for $3.00? (9 granola bars)

◐◐○○○ If you had 3 quarters, 4 nickels, and 5 pennies, how many granola
bars could you buy? (3 granola bars) If I had 5 dimes, 8 nickels, and
45 pennies, how many granola bars could I buy? (4 granola bars)

◐◐◐○○ About how much would 1 granola bar cost? Round your answer
up to the nearest penny. (34¢) About how much would 2 granola
bars cost? (67¢)

Operations and Algebraic Thinking

Dinosaur Sizes* | Number Stories

Triceratops, a dinosaur with 3 horns, was about 8 to 10 feet tall and about 20 to 25 feet long.

◉○○○○ Why would scientists measure the height and length of *Triceratops?* Would we measure both of these for a person? (No.) For a horse? (Yes.) What are some possible heights and lengths for *Triceratops?* (8 ft tall, 20 ft long; 9 ft tall, 21 ft long; and so on)

◉◉○○○ How many horns do 2 *Triceratops* have? (6 horns) 3 *Triceratops?* (9 horns)

◉◉◉○○ If *Triceratops* was about 8 feet tall and about 20 feet long, how much longer was it than it was tall? (12 ft)

◉◉◉◉○ If *Triceratops* was 8 feet tall and 24 feet long, how many times longer was it than it was tall? (3 times longer)

Variation Ask about *Stegosaurus* (12–13 feet tall and 18–25 feet long).

*You may want to display the information about *Triceratops.*

Double, Triple, Quadruple

Say: *The number is **5. Double it.** (10)*

⬤〇〇〇〇 Choose a 1-digit number. Tell children to double it.

⬤⬤〇〇〇 Choose a 1-digit number. Tell children to triple it.

⬤⬤⬤〇〇 Choose a 1-digit number. Tell children to quadruple it.

⬤⬤⬤⬤〇 Choose a 2-digit number. Tell children to triple or quadruple it.

Age: People vs. Mice Number Stories

Many mice live to be 4 years old. Many people live to be 80 years old.

●○○○○ Are there many mice as old as you? (No.) Are there many people as old as you? (Yes.) How many birthdays do most mice have? (4 birthdays) How many birthdays have you had?

●●○○○ About how many years longer than a mouse do many people live? (76 years) If you live to be 80 years old, about how many more years will you live?

●●●○○ How long is half of a 4-year-old mouse's life? (2 years) About how long is half of an 80-year-old human's life? (40 years) About how long is half of your life so far?

●●●●○ About how many times as long as a mouse do many people live? (20 times as long)

A Dog's Age Number Stories

One rule of thumb for finding a dog's equivalent "human age" says to count 7 human years for each year of the dog's life.

● ○ ○ ○ ○ If a dog is about 1 year old, what is its human age using this rule? (7 years old) If a dog is about 2 years old, what is its human age? (14 years old)

● ● ○ ○ ○ If a dog's equivalent human age is 21 years, about how many years has it been alive? (3 years) If its equivalent human age is 49 years, about how many years has the dog been alive? (7 years)

● ● ● ○ ○ Using "times" or "multiply," restate this rule of thumb. (Multiply the dog's age by 7 to find its equivalent human age.)

● ● ● ● ○ Which do you think usually live longer, dogs or humans? (Humans) How can you tell by this rule of thumb? (Each year of the dog's life counts for several years of human life.)

Another Dog's Age Number Stories

A new rule of thumb for finding the equivalent "human age" of a dog takes into account that a dog can have puppies at 1 year and is full-grown by 2 years. The rule suggests counting the first year of a dog's life as 15 years of human life, the second as 10 more, and each year after as 5 more years.

●○○○○ Using the new rule, what is the human age of a 1-year-old dog? (15 years old) Of a 4-year-old dog? (35 years old)

●●○○○ Using the new rule, about how many years has a dog been alive if its equivalent human age is 50? (7 years) If its equivalent human age is 65? (10 years)

●●●○○ Consider the old rule of thumb that said, "To find a dog's equivalent human age, multiply the dog's actual age by 7." Are the old and new estimates of the dog's equivalent human age closer when the dog is actually 1 year old, 5 years old, or 10 years old? (At the actual age of 1, the estimates are 8 years apart; at the actual ages of 5 and 10, the estimates are 5 years apart.)

Whale Speeds `Number Stories`

The sperm whale travels from 3 to 5 miles per hour if it is left undisturbed. When it is in danger, the sperm whale can swim at a speed of more than 13 miles per hour.

◉○○○○ If a sperm whale is swimming undisturbed, is it likely to swim more than 10 miles in 2 hours? (No.)

◉◉○○○ About how many miles per hour faster does a sperm whale travel when it is in danger than when it is left undisturbed? (8–10 mi per hr faster)

◉◉◉○○ If a sperm whale is left undisturbed, about how many miles will it swim in 3 hours? (9–15 mi) If it is in danger, about how many miles might it travel in the same amount of time? (More than 39 mi)

Operations and Algebraic Thinking

Baseball Speeds   Number Stories

In baseball, a type of batted ball that flies low and fast, and usually in a straight line, is called a line drive. A line drive travels about 100 yards in 4 seconds.

●○○○○   About how many yards does a line drive travel in 2 seconds? (50 yd) In 1 second? (25 yd)

●●○○○   About how many feet does a line drive travel in 4 seconds? (300 ft) About how many inches does a line drive travel in 4 seconds? (3,600 in.)

Numbers After and Before

Say: *I will point to three of you, one after the other. The first child will say a number such as* **16.** *The second says the number after* **16,** *and the third says the number before* **16.** *Then we will do it again.*

◉○○○○ Use numbers 1–100.

◉◉○○○ Use numbers 100–1,000.

◉◉◉○○ Use numbers greater than 1,000.

◉◉◉◉○ Use negative numbers.

Whole Numbers Between

Ask: *What series of whole numbers comes between the numbers **12** and **15**?*
Also reword the question as: *What numbers are greater than **23** and less than **28**?*

●○○○○ Choose pairs of whole numbers 1–50 within the same decade.
Examples: 20 and 25 or 19 and 14.

●●○○○ Choose pairs of whole numbers 1–100 that cross decades.
Examples: 64 and 56 or 73 and 68.

●●●○○ Choose pairs of whole numbers 100–5,000 that cross decades or
centuries. Examples: 397 and 389 or 397 and 404.

●●●●○ Specify dollars or cents. Example: *How many more cents is 64¢
than 53¢?* (11¢)

Variation Say: *Name every other number between 30 and 40.*
(32, 34, 36, 38)

The Oldest Living Animal   Number Stories

The oldest living animal on record was found in 2006. At that time, this clam was 507 years old. (It was an ocean quahog, *Arctica islandica*.)

◐○○○○   If the clam were still alive today, how old would it be?

◐◐○○○   Was the clam alive in 1906? (Yes.) Was it alive in 1806? (Yes.) In 1706? (Yes.) In 1606? (Yes.) In 1506? (Yes.) In 1406? (No.)

◐◐◐○○   How old was the clam in 1706? (207 years old) In 1606? (107 years old) In 1506? (7 years old) In 1500? (1 year old)

◐◐◐◐○   When was the clam hatched? (Probably in 1499) In what year would the clam be 600 years old if it lives that long? (2099)

Number and Operations in Base Ten

Outdoor Temperatures Number Stories

It is about _____ outside right now. (Use a current thermometer reading. Sometimes use °C, sometimes °F.)

●○○○○ If the high temperature yesterday was about 60°F, is today warmer or cooler than yesterday? About how much warmer or cooler?

●●○○○ If the expected high today is 73°F, about how many more degrees will the temperature have to rise to reach the expected high? If the predicted temperature for tonight is 58°F, about how many degrees will the temperature change from what it is now?

●●●○○ If the temperature in an Arizona desert is usually about 40°F warmer than the temperature here, what is the approximate temperature in that desert right now?

●●●●○ If the coldest day this winter was about −5°F, about how much warmer is today than the coldest day of the year?

Missing Numbers

Say: *Listen carefully and tell me which numbers I am missing:*
40, 35, 25, 20. (30)

●○○○○ Count by 1s using numbers 1–50. Example: *21, 22, 24, 25.* (23)

●●○○○ Count by 2s, 5s, or 10s using numbers 1–50.
Example: *10, 15, 20, 30, 35.* (25)

●●●○○ Count backward by 1s, 2s, 5s, or 10s using numbers 1–100.
Example: *87, 86, 85, 83.* (84)

●●●●○ Count forward or backward by 1s, 2s, 5s, 10s, or 100s using
numbers 100–1,000. Example: *332, 334, 336, 338, 342.* (340)

How Many?

Ask: *If there are **50 ears,** how many **people** are there?* (25 people)

⦿○○○○ Choose things that come in pairs, such as ears, hands, feet, and arms.

⦿⦿○○○ Choose things that come in 10s, such as fingers and toes.

⦿⦿⦿○○ Choose things that come in 4s, such as chair legs and table legs.

Elephant Meals Number Stories

Elephants spend about 16 hours of every day eating.

⬤○○○○ About how many hours per day do elephants **not** eat? (8 hr) Do elephants spend more of the day eating or not eating? (Eating) About how many hours per day do you usually spend eating? How many hours per day do you **not** eat?

⬤⬤○○○ About how many hours do elephants spend eating in 2 days? (32 hr) In 3 days? (48 hr)

⬤⬤⬤○○ On average, what fraction, or portion, of the day do elephants spend eating? $\left(\frac{16}{24} \text{ or } \frac{2}{3}\right)$ What fraction, or portion, of the day do they spend **not** eating? $\left(\frac{8}{24} \text{ or } \frac{1}{3}\right)$

⬤⬤⬤⬤○ About how many minutes per day do elephants spend doing things other than eating? (480 min)

New Books Number Stories

Every day, about 800 new books are published in the United States.
About 60 of these books are children's books.

●○○○○ About how many books for adults are published each day?
(740 books)

●●○○○ About how many children's books are published in 2 days?
(120 children's books) In 3 days? (180 children's books) What is the
average number of books published in 2 days? (1,600 books)
In 3 days? (2,400 books)

●●●○○ About how many children's books are published in 5 days?
(300 children's books) In 15 days? (900 children's books)

●●●●○ About how many days would it take to publish 1,200 children's
books? (20 days)

Continue the Sequence

Say: *Raise your hands as I begin to count. I will point to each of you, one at a time. When I point to you, say the next number in the sequence and put down your hand. **5, 10, 15,** . . .*

After each child has a number, say: *Let's line up in number order from largest number to smallest* (or *smallest number to largest*).

- ●○○○○ Count by 1s, 2s, 5s, or 10s. Begin at any multiple of that number. Example: *35, 40, 45.* (*50, 55, 60,* . . .)

- ●●○○○ Count backward by 1s, 2s, 5s, or 10s.

- ●●●○○ Count forward or backward using numbers such as 3, 4, 6, or 7. Example: *4, 8, 12.* (*16, 20, 24,* . . .)

- ●●●●○ Count forward or backward using sequences of fractions or decimals. Example: $\frac{1}{4}, \frac{2}{4} \left(\frac{1}{2}\right), \frac{3}{4}, 1, 1\frac{1}{4}.$ $\left(1\frac{2}{4}, 1\frac{3}{4}, 2, \dots\right)$

CCSS 1.NBT.1, 2.NBT.2, 3.NF.1   75

Repeated Digits

Say: *Name a number that is written with* **two** *7s in it.* Be sure children say the number correctly.

◐○○○○ Use two repeated digits. Examples: 77; 277; 747; 7,736

◐◐○○○ Use three repeated digits. Examples: 777; 3,777; 7,277; 867,757

◐◐●○○ Use four repeated digits. Examples: 7,777; 477,377

Variation Repeat the exercise with a number other than 7.

Tall Buildings* Number Stories

The Willis Tower is about 443 meters tall. The John Hancock Center is about 343 meters tall. The Eiffel Tower is about 300 meters tall.

●○○○○ Which of these buildings is the tallest? (Willis Tower) Which is the shortest? (Eiffel Tower)

●●○○○ About how much taller than the Eiffel Tower is the Willis Tower? (143 m) About how much taller than the John Hancock Center is the Willis Tower? (100 m) About how much shorter than the John Hancock Center is the Eiffel Tower? (43 m)

●●●○○ About how many centimeters tall is the Willis Tower? (44,300 cm) One decimeter equals 10 centimeters. How many decimeters tall is the Willis Tower? (4,430 dm)

Variation Pose questions using the heights of some other tall buildings: Great Pyramid of Cheops, 147 m; Empire State Building, 381 m; Aon Center (Chicago), 346 m.

*You may want to display the heights of the buildings.

CCSS **2.NBT.4, 2.MD.5, 3.NBT.2**

Number and Operations in Base Ten

Breathing Number Stories

People take approximately 12 breaths per minute when they are relaxed.

⦿○○○○ About how many breaths do people take in 2 minutes if they are relaxed? (24 breaths) In 3 minutes? (36 breaths) In 4 minutes? (48 breaths)

⦿⦿○○○ About how many breaths do people take in half an hour if they are relaxed? (360 breaths) In 1 hour? (720 breaths)

⦿⦿⦿○○ During physical education class, will people take more or less than 240 breaths in 20 minutes? Explain. (More than 240 breaths; they would take about 240 breaths in 20 minutes if they were relaxed. When they are active, they take more breaths per minute.)

Variation Ask children to count how many breaths they take while you time them for one minute. Help them find the middle value of the class results, and then ask questions such as those above.

Numbers with *n* Digits

Ask: *Which is the smallest **1-digit** number? Which is the largest? How many **1-digit** numbers are there?*

⦿○○○○ Use 1-digit numbers. (0 is smallest; 9 is largest; there are 10 numbers.) These are the ones numbers.

⦿⦿○○○ Use 2-digit numbers. (10 is smallest; 99 is largest; there are 90 numbers.) These are the tens.

⦿⦿⦿○○ Use 3-digit numbers. (100 is smallest; 999 is largest; there are 900 numbers.) These are the hundreds.

⦿⦿⦿⦿○ Use 4-digit numbers. (1,000 is smallest; 9,999 is largest; there are 9,000 numbers.) These are the thousands.

⦿⦿⦿⦿⦿ Use 5-digit numbers. (10,000 is smallest; 99,999 is largest; there are 90,000 numbers.) These are the ten-thousands.

Although numerals such as 01 or 0032 have not been counted here, commend children who consider these possibilities.

(CCSS) 1.NBT.2, 1.NBT.3, 2.NBT.1, 2.NBT.4   79

Easier Numbers

Say: *Change the number **19** to the nearest easy number that ends in **0**.* (20)

●○○○○ Use 2-digit numbers and provide children with two choices for their response. Example: *Is the number 27 closer to 20 or to 30?* (30)

●●○○○ Use 2-digit numbers. Example: *Change 27 to the nearest easy number that ends in 0.* (30)

●●●○○ Use 3-digit numbers. Example: *Change 384 to the nearest easy number that ends in 0.* (380)

●●●●○ Use 3- or 4-digit numbers. Tell children to change the numbers to the nearest 100. Examples: *Change 364 to the nearest 100.* (400) *Change 4,692 to the nearest 100.* (4,700)

Variation Display numbers. Ask similar questions.

Bird Egg Sizes   Number Stories

Ostrich eggs are about 6 to 8 inches long. Hummingbird eggs are less than $\frac{1}{2}$ inch long.

- ●○○○○ Which is longer, an ostrich egg or a hummingbird egg? (An ostrich egg)

- ●●○○○ How much longer than a half-inch long hummingbird egg is a 6-inch long ostrich egg? $\left(5\frac{1}{2}\text{ in. longer}\right)$ How much shorter than an 8-inch long ostrich egg is a half-inch long hummingbird egg? $\left(7\frac{1}{2}\text{ in. shorter}\right)$

- ●●●○○ How many half-inch hummingbird eggs placed end to end are as long as an 8-inch long ostrich egg? (16 hummingbird eggs)

- ●●●●○ How many times longer than a half-inch hummingbird egg is a 6-inch long ostrich egg? (12 times longer) How many times longer than a half-inch hummingbird egg is an 8-inch long ostrich egg? (16 times longer)

Number and Operations in Base Ten

CCSS 3.NBT.2, 3.NF.1   81

Hummingbird Wing Flaps   Number Stories

A small hummingbird can beat its wings about 70 times per second.

◐○○○○   How many times can you flap your arms in 1 second?
(Maybe 1 time)

◐◐○○○   About how many times can a hummingbird beat its wings in
2 seconds? (140 times) In 3 seconds? (210 times) About how
many times can a hummingbird beat its wings in half a second?
(35 times)

◐◐◐○○   About how many times can a hummingbird beat its wings in
1 minute? (4,200 times) In 2 minutes? (8,400 times)

Creating Numbers

Say: *Use the digits **4, 3,** and **7** to create a number.* (Choose and adjust digits and the number of digits.)

◉○○○○   Give children digits to create the largest 2-digit number; the smallest 2-digit number; the largest 3-digit number; the smallest 3-digit number; and so on.

◉◉○○○   Give children digits to create a number with the 3 in the hundreds place; with 3 in the tens place; with 3 in the ones place; and so on.

◉◉◉○○   Give children digits to create any number with 2 decimal places; the largest number with 2 decimal places; the smallest number with 2 decimal places; and so on.

◉◉◉◉○   Give children digits to create a fraction less than 1; greater than 1; greater than $\frac{1}{2}$; less than $\frac{1}{2}$; and so on.

◉◉◉◉◉   Give children digits to create the largest fraction they can or to create the smallest fraction they can.

Number and Operations in Base Ten

Guess My Number

Say: *I'm thinking of a number greater than **55** but less than **58**. What might the number be?*

or

Say: *I'm thinking of a number less than **72** but greater than **68**. What might the number be?*

●○○○○ Use a pair of whole numbers 1–100.

●●○○○ Use a pair of whole numbers 100–1,000.

●●●○○ Use a pair of fractions or decimals.

●●●●○ Use a pair of negative integers.

Remember, there are an infinite number of possibilities between any two numbers. Commend children who suggest fractions or decimals.

Extreme Temperatures Number Stories

As of 2013, the hottest temperature ever recorded on Earth was about 136°F; this was in Libya. The coldest temperature ever recorded for a continuously inhabited area outside Antarctica was about −90°F; this was in Russia.

●○○○○ Is −90°F warmer or colder than −130°F? (Warmer) Is −90°F warmer or colder than −80°F? (Colder)

●●○○○ About how many degrees Fahrenheit is it today? About how many degrees warmer was the hottest day in Libya?

●●●○○ About how many degrees hotter was the hottest day in Libya than the coldest day in Russia? (226°F)

Variation Ask similar questions using Celsius temperatures. The hottest temperature was 58°C; the coldest was −68°C.

CCSS 3.NBT.2   85

Alligator Eggs Number Stories

A female alligator lays an average of 40 eggs each time she has young.

◉○○○○ Do some female alligators lay more than 40 eggs when they have young? (Yes.) How do you know? (40 eggs is an average; that means some females lay more eggs than the average and some lay less.)

◉◉○○○ Is the average number of eggs laid by a female alligator more or less than 1 dozen? (More) More or less than 2 dozen? (More) More or less than 3 dozen? (More) More or less than 4 dozen? (Less) How many less? (8 less)

◉◉◉○○ What is the average number of eggs laid by 2 female alligators? (80 eggs) By 3 female alligators? (120 eggs)

Adding and Subtracting Multiples of 10

Ask children to add or subtract to solve problems involving multiples of 10.
For example: $35 + 10 = ?$ (45) and $50 - 10 = ?$ (40)

- ●○○○○ Add or subtract 10 to or from numbers divisible by 5 or 10.
  Examples: $30 + 10 = ?$ (40) and $25 - 10 = ?$ (15)

- ●●○○○ Add or subtract 10 to or from any 2-digit numbers.
  Examples: $17 + 10 = ?$ (27) and $63 - 10 = ?$ (53)

- ●●●○○ Add or subtract 100 or 1,000 to or from any number.
  Examples: $23 + 100 = ?$ (123) and $223 + 100 = ?$ (323)

- ●●●●○ Add or subtract multiples of 10, 100, or 1,000 to or from any
  number. Examples: $453 + 40 = ?$ (493) and $5,000 + 898 = ?$ (5,898)

Multiplication Properties of 10

Ask children to multiply to solve problems involving multiples of 10.
For example: $3 \times 10 = ?$ (30) and $23 \times 100 = ?$ (2,300)

- ●○○○○ Use multiples of dimes. Example: *How much money is 3 dimes?* (30¢)

- ●●○○○ Use multiples of 10. Example: *How much is four 10s?* (40)

- ●●●○○ Multiply 1- or 2-digit numbers by 10.
  Examples: $22 \times 10 = ?$ (220) and $34 \times 10 = ?$ (340)

- ●●●●○ Multiply 3- or 4-digit numbers by 10, 100, or 1,000.
  Examples: $342 \times 10 = ?$ (3,420) and $467 \times 100 = ?$ (46,700)

Airplane Speed vs. Human Speed Number Stories

Jets travel an average of about 500 miles per hour. People walk an average of about 3 miles per hour.

●○○○○ About how many miles could an airplane travel in 3 hours? (1,500 mi) About how far could a person walk in 3 hours? (9 mi)

●●○○○ About how many miles per hour faster do airplanes travel than people who are walking? (497 mi per hr faster) About how many miles farther could an airplane travel in 3 hours than a person walking? (1,491 mi farther)

●●●○○ About how many miles could an airplane travel in 8 hours? (4,000 mi) About how many miles could a person walk in 8 hours? (The mathematical answer is 24 miles, but most people could not continue to walk 3 miles per hour for that long.)

Bacteria Growth Number Stories

Bacteria are tiny living creatures that can only be seen with a microscope. If bacteria are given food and a warm, wet place in which to grow, they can double their number about every half hour.

●○○○○ If 100 bacteria live under these conditions, about how many will there be in half an hour? (200 bacteria) In an hour? (400 bacteria) In 2 hours? (1,600 bacteria)

●●○○○ If there are 60 of these tiny creatures, about how many were there half an hour ago? (30 bacteria) 1 hour ago? (15 bacteria)

●●●○○ If there are 10 bacteria, about how long will it be until there are 160 bacteria? (2 hr) If there are 20 bacteria, about how long will it be until there are 160 bacteria? $\left(1\frac{1}{2}\text{ hr}\right)$

Siblings

Ask: *How many of you have a brother? How many have a sister?*

○○○○○ Ask: *Are there more of you with brothers or more with sisters? How many more? There are **25** children in this class, yet only **22** hands in all were raised for both questions. Why?* (If the total number of hands is fewer than the number of children in class, some do not have siblings. If more, some have both brothers and sisters.)

●●○○○ Say: *More of you have brothers than sisters. Does that mean that the total number of boy children in your families is more than the total number of girl children?* (No. All with sisters may have many sisters.) *How could we find the total numbers of brothers and sisters?* (Ask how many brothers and how many sisters children have.)

●●●○○ Ask: *What fraction of the people in our class have brothers? Sisters?* Ask the boys with sisters: *Do all your sisters have a brother?* (Some children answer no—forgetting themselves—if there is not another brother.) Ask a similar question of the girls.

CCSS 1.NBT.3, 1.NBT.4, 2.NBT.5, 3.NBT.2, 3.NF.1

Estimating Differences

Ask: *To the nearest **10**, about what is the difference between **34** and **82**?*
(About 50)

⬤○○○○ Have children estimate the difference to the nearest 10 between two 2-digit numbers.

⬤⬤○○○ Have children estimate the difference to the nearest 100 between two 3-digit numbers. Example: *To the nearest 100, what is the difference between 292 and 586?* (About 300)

⬤⬤⬤○○ Have children estimate the difference to the nearest 1,000 between two 4-digit numbers. Example: *To the nearest 1,000, what is the difference between 2,535 and 8,624?* (About 6,000)

Popping Corn Number Stories

Typically, you get approximately 34 cups of popcorn from 1 cup of kernels.

●○○○○ Why does the popcorn take up more space than the kernels?
(When kernels pop, they expand.)

●●○○○ About how many cups of popcorn will you get if you pop $\frac{1}{2}$ cup of
kernels? (17 cups) If you pop about 2 cups of kernels? (68 cups)

●●●○○ If you wanted about 100 cups of popcorn, how many cups of
kernels should you pop? (About 3 cups, or a little less than 3 cups)

●●●●○ About how many people can you feed if you pop 1 cup of kernels?
(Insufficient data to answer the question; children may want to
estimate anyway.)

Earth's Orbit   Number Stories

Earth travels about 18 miles per second in its orbit (path) around the sun.

⦿○○○○ Does Earth travel more or less than 18 miles in one minute?
(More) Does Earth travel more or less than 18 feet in one second?
(More)

⦿⦿○○○ About how far does Earth travel in 2 seconds? (36 mi)
In 3 seconds? (54 mi)

⦿⦿⦿○○ Suppose Earth has moved 80 miles. Have more or less than
5 seconds passed? (Less) Suppose Earth has moved 100 miles.
Have more or less than 3 seconds passed? (More)

Parts in a Whole

Ask children to name fractional parts.

●○○○○ *How many **thirds** in 1?* (3)

●●○○○ *How many **thirds** in 2?* (6) *In 3?* (9)

●●●○○ *How many **thirds** in 10?* (30) *In 100?* (300)

Variation Substitute any other fraction for thirds.

Body Muscle Number Stories

Less than half of the average person's body weight is muscle.

●○○○○ How much of a person's body weight is not muscle? $\left(\text{More than } \frac{1}{2}\right)$

●●○○○ If a child weighs about 70 pounds, about how much do the child's muscles weigh? (Less than 35 lb)

To be more exact, about $\frac{2}{5}$ of a person's body weight is muscle.

●●●○○ What fraction of a person's body weight is not muscle? $\left(\text{About } \frac{3}{5}\right)$

●●●●○ For a 100-pound person, about how many pounds are muscle? (40 lb) About how many pounds are not muscle? (60 lb)

●●●●● About what percent of a person's weight is muscle? (40 percent) About what percent is not muscle? (60 percent)

Making Orange Juice   Number Stories

As a rule of thumb, an orange will yield about a half cup of orange juice.

●○○○○ How many oranges would you need to make about 1 cup of orange juice? (2 oranges)

●●○○○ How many oranges would you need to make about 2 cups of orange juice? (4 oranges) To make about 4 cups of orange juice? (8 oranges) To make about a cup of orange juice for each member of your family?

●●●○○ If we have 10 oranges, about how many cups of orange juice can we make? (5 cups) If we have 40 oranges? (20 cups) If we have 100 oranges? (50 cups)

●●●●○ There are 2 cups in 1 pint. How many oranges would you need to make about a pint of orange juice? (4 oranges) There are 4 cups in 1 quart. How many oranges would you need to make about a quart of orange juice? (8 oranges)

CCSS 2.G.3, 3.OA.7, 3.NF.1, 3.MD.2   97

Number and Operations — Fractions

Parts

Ask: *What is $\frac{1}{2}$ of **10**?* (5)

⚫⚪⚪⚪⚪ Have children determine $\frac{1}{2}$ of any 1-digit even number.

⚫⚫⚪⚪⚪ Have children determine $\frac{1}{2}$ of any 2-digit even number.

⚫⚫⚫⚪⚪ Have children determine $\frac{1}{2}$ of any 3-digit even number.

⚫⚫⚫⚫⚪ Have children calculate $\frac{1}{3}$ or $\frac{1}{4}$ of 2-digit numbers divisible by 3 or 4.
Examples: *What is $\frac{1}{3}$ of 27?* (9) *What is $\frac{1}{4}$ of 16?* (4)

⚫⚫⚫⚫⚫ Have children estimate $\frac{1}{3}$, $\frac{1}{4}$, or $\frac{1}{10}$ of 2- or 3-digit numbers.
Example: *About how much is $\frac{1}{3}$ of 71?* (About 23 or 24)

Body Water Weight Number Stories

About half a person's weight is water.

⚫○○○○ About how much of a person's weight is **not** water? (Half)

⚫⚫○○○ Kim weighs 60 pounds. About how much does the water in his body weigh? (30 lb) Lisa weighs 50 pounds. About how much does the water in her body weigh? (25 lb)

⚫⚫⚫○○ About what percent of a person's weight is water? (50 percent) About what percent of a person's weight is **not** water? (50 percent)

Making Pancakes Number Stories

About $\frac{1}{4}$ cup of pancake batter makes one round pancake 4 inches across.

●○○○○ Will $\frac{1}{2}$ cup of batter make a round pancake that is larger or smaller than 4 inches across? (Larger) Will $\frac{1}{8}$ cup of pancake batter make a round pancake that is larger or smaller than 4 inches across? (Smaller)

●●○○○ How many 4-inch round pancakes could you make with 1 cup of batter? (4 pancakes) With 2 cups of batter? (8 pancakes) With 3 cups of batter? (12 pancakes)

●●●○○ There are 2 cups in 1 pint. Using 1 pint of batter, how many 4-inch pancakes could you make? (8 pancakes) Using $1\frac{1}{2}$ pints of batter? (12 pancakes)

●●●●○ If each member of your family wanted 4 pancakes, how many cups of batter would you need? If each member of this class wanted 2 pancakes, how much batter would you need?

Measurement Fractions

Ask: *What part of an **hour** is **45 minutes?*** $\left(\frac{45}{60}\text{ or }\frac{3}{4}\text{ of an hour}\right)$

(Each of the following variations can be made more or less difficult by altering the size of the numbers and how easily they divide into desired unit amounts)

Variation 1 Ask about measures of time: *What part of an hour is 30 minutes?* $\left(\frac{30}{60}\text{ or }\frac{1}{2}\text{ of an hour}\right)$

Variation 2 Ask about measures of money: *What part of a dollar is 50¢?* $\left(\frac{50}{100}\text{ or }\frac{1}{2}\text{ of a dollar}\right)$ *Two dimes?* $\left(\frac{20}{100},\frac{2}{10},\text{ or }\frac{1}{5}\text{ of a dollar}\right)$

Variation 3 Ask about measures of distance, capacity, or weight: *A pound is 16 ounces. What part of a pound is 8 ounces?* $\left(\frac{1}{2}\text{ lb}\right)$ *4 ounces?* $\left(\frac{4}{16}\text{ or }\frac{1}{4}\text{ lb}\right)$

Koala Birth Weights Number Stories

A newborn koala weighs about $\frac{1}{10}$ of an ounce when it is born. This is about as heavy as one penny.

●○○○○ You have 2 pennies in one hand. How many newborn koalas would you have to put in the other hand to have about the same weight in each hand? (2 koalas) If you added another newborn koala to one hand, do you think that you could feel the difference? (Probably not.)

●●○○○ About how many newborn koalas would weigh about 1 ounce? (10 koalas) About how many pennies? (10 pennies)

●●●○○ About how many newborn koalas would weigh about 2 ounces? (20 koalas) There are 16 ounces in 1 pound. About how many newborn koalas would weigh about 1 pound? (160 koalas)

A Pint Is a Pound Number Stories

For water, it is said, "A pint is a pound the world around."

● ○ ○ ○ ○   If we had 2 pounds of water, about how many pints would we have? (2 pt, or 1 qt) If we had 3 pounds of water, about how many pints would we have? (3 pt)

● ● ○ ○ ○   There are 2 pints in 1 quart. About how much would 1 quart of water weigh? (2 lb) About how much would 2 quarts of water weigh? (4 lb)

● ● ● ○ ○   There are 8 pints in 1 gallon. About how much would a gallon of water weigh? (8 lb) About how much would 3 gallons of water weigh? (24 lb)

● ● ● ● ○   There are 2 cups in a pint. About how much would 1 cup of water weigh? $\left(\frac{1}{2}\ \text{lb}\right)$ Half a cup of water? $\left(\frac{1}{4}\ \text{lb}\right)$

Number and Operations — Fractions

Water in an Elephant Trunk   Number Stories

The trunk of an average elephant can hold about $1\frac{1}{2}$ gallons of water.

●○○○○ About how many 1-gallon milk cartons of water can the average elephant hold in its trunk? $\left(1\frac{1}{2}\text{ cartons}\right)$

●●○○○ About how much water can the trunks of 2 average elephants hold? (3 gal) Of 3 average elephants? $\left(4\frac{1}{2}\text{ gal}\right)$

●●●○○ How many average elephants would be needed to hold about 3 gallons of water? (2 elephants) To hold about 6 gallons of water? (4 elephants)

●●●●○ There are 4 quarts in 1 gallon. About how many quarts of water can an average elephant hold in its trunk? (6 qt) There are 8 pints in 1 gallon. About how many pints of water can an average elephant hold in its trunk? (12 pt)

Money and Measure Counts

Say: *Let's count nickels from **15¢: 15¢, 20¢, 25¢,** . . .*

⦿○○○○ Count nickels or dimes forward and backward from any amount.

⦿⦿○○○ Count quarters or half-dollars forward and backward from any amount.

⦿⦿⦿○○ Count $\frac{1}{4}$ cups, $\frac{1}{2}$ cups, or $\frac{1}{3}$ cups forward and backward from any amount. Example: *Let's count $\frac{1}{2}$ cups from $\frac{1}{2}$ cup: $\frac{1}{2}$ cup, 1 cup, $1\frac{1}{2}$ cups, 2 cups,* . . .

Measuring Tools

Say: *I'm thinking of a measuring tool that can measure **amounts of time**. What is the tool?* (Clock or calendar) Also point out informal tools like the position of the sun and moon.

●○○○○ Think of a measuring tool that can measure length. (Ruler, tape measure, and meterstick or yardstick) Also point out informal measures like a person's foot or a piece of paper.

●●○○○ Think of a measuring tool that can measure mass. (Pan balance) Think of a tool that can measure capacity. (Liter beaker and measuring cups marked in milliliters and liters)

Informal Measuring Tools

Ask: *What could we use to measure the **length** of this **rug** if we had no **rulers** or **metersticks** or **tape measures?*** (A foot, a piece of paper, the length from elbow to fingertip, and so on) *What are the advantages and disadvantages of measuring this way?*

●○○○○ Ask: *What could we use to measure the length of our thumbs? This book? This room? A school bus?*

●●○○○ Ask: *What could we use to measure time?* (The sun, the moon, and so on) *What could we use to measure temperature?* (The clothes that we need to wear in order to feel cool or warm enough tell approximately how warm or cold it is.)

●●●○○ Ask: *What could we use to measure weight?* (Holding things in our hands, putting things in water to see how quickly they sink, and so on)

Making a Dollar

Ask: *How many **pennies** do we need to make $1.00?* (100 pennies)

●○○○○ Ask: *How many dimes do we need to make $1.00?* (10 dimes)
How many nickels do we need to make $1.00? (20 nickels)

●●○○○ Ask: *How many quarters do we need to make $1.00?* (4 quarters)
How many quarters do we need to make $3.00? (12 quarters)

●●●○○ Ask how many more of a given coin are needed to make $1.00 if
we have a given amount of the same coin. Example: *How many
more quarters do we need to make $1.00 if we have 1 quarter?*
(3 more quarters)

●●●●○ Ask how many more of a given coin are needed to make $1.00 if
we have a given amount of different coins. Example: *How many
more nickels do we need to make $1.00 if we have 3 dimes and
5 pennies?* (13 more nickels)

Money Exchanges

Ask: *How many **nickels** can we get if we have **31 pennies?*** (6 nickels)

⬤○○○○ Exchange pennies for nickels or dimes. Example: *How many dimes can we get if we have 46 pennies?* (4 dimes)

⬤⬤○○○ Exchange one type of coin (pennies, nickels, dimes, or quarters) for another. Example: *How many quarters can we get if we have 6 dimes?* (2 quarters)

⬤⬤⬤○○ Exchange one-dollar bills for ten-dollar bills. Example: *How many ten-dollar bills can we get if we have $16?* (1 ten-dollar bill)

⬤⬤⬤⬤○ Exchange ten-dollar bills for hundred-dollar bills. Example: *How many hundred-dollar bills can we get if we have 14 ten-dollar bills?* (1 hundred-dollar bill)

More Money Exchanges

Ask: *If I trade **4 dimes** for **46 pennies**, am I getting a good deal?*
(Yes. I give 40¢; I receive 46¢.)

●○○○○ Exchange pennies for nickels or dimes. Example: *If I trade
9 pennies for 2 nickels, am I getting a good deal?* (Yes. I give 9¢;
I receive 10¢.)

●●○○○ Exchange one type of coin (pennies, nickels, dimes, or quarters)
for another. Example: *If I trade 8 nickels for 3 dimes, am I getting a
good deal?* (No. I give 40¢; I receive 30¢.)

●●●○○ Exchange one-dollar bills for five-dollar or ten-dollar bills.
Example: *If I trade 14 one-dollar bills for 2 ten-dollar bills, am
I getting a good deal?* (Yes. I give $14; I receive $20.)

●●●●○ Exchange ten-dollar bills for fifty-dollar or hundred-dollar bills.
Example: *If I trade 12 ten-dollar bills for 2 hundred-dollar bills,
am I getting a good deal?* (Yes. I give $120; I receive $200.)

The Length of a Dollar Bill   Number Stories

A dollar bill is about 6 inches long.

○●○○○○ About how long is a ten-dollar bill? (About 6 in. long) Is a dollar bill more or less than 6 inches wide? (Less)

●●○○○ About what part, or fraction, of a foot is the length of a dollar bill? $\left(\text{About } \frac{6}{12} \text{ or } \frac{1}{2} \text{ ft}\right)$ About how many dollar bills would you need to put end to end to make 1 foot? (2 dollar bills) About how many dollar bills would you need to put end to end to make 2 feet? (4 dollar bills)

●●●○○ About how many dollar bills would you need to put end to end to make 1 yard? (6 dollar bills) About how many dollar bills would you need to put end to end to make 2 yards? (12 dollar bills)

Koala Bears Number Stories

At birth, a koala is about 2 centimeters long. By the time it comes out of its mother's pouch, it is about 20 centimeters long.

○○○○○ Show me, with your hands, about how big the koala is when it is born. Show me, with your hands, about how big the koala is when it comes out of its mother's pouch.

●●○○○ How much does a koala grow before it comes out of the pouch? (18 cm)

●●●○○ How many times bigger is the koala when it comes out of its mother's pouch than when it is born? (10 times bigger)

Coin Equivalents

Ask: *If I have 24¢, what coins might I have?* (Possibilities include: 2 dimes and 4 pennies; 4 nickels and 4 pennies; and so on.)

●○○○○ Use amounts under $1. Example: *If I have 47¢, what coins might I have?* (Possibilities include: 4 dimes and 7 pennies; 2 dimes, 5 nickels, and 2 pennies; and so on.)

●●○○○ Use amounts over $1. Example: *If I have $1.13, what coins might I have?* (Possibilities include: 4 quarters, 1 dime, and 3 pennies; 11 dimes and 3 pennies; and so on.)

●●●○○ Use amounts under $1 and ask for the smallest number of coins possible. Example: *What is the smallest number of coins I could use to make 36¢?* (3 coins: 1 quarter, 1 dime, and 1 penny)

Money and Time

Ask: *Would you rather have **a penny per day for a month**, or **a nickel per day for a week?*** (About 30¢ versus 35¢) (Choose different coin amounts per calendar period. Increase the complexity by increasing the difficulty of the multiplication necessary to determine the answer.)

●○○○○ *A dime per day for a month or a penny per day for a year?*
(About $3.00 versus $3.65)

●●○○○ *A nickel per day for 3 months or a penny per day for a year?*
(About $4.50 versus $3.65)

●●●○○ *A quarter per month for 2 years or a dime per day for 2 months?*
(Both equal about $6.00; if the months are 31 days each, the dimes would equal $6.20.)

How Many?

Ask: *How many **dimes** and **pennies** in **31¢?*** (3 dimes and 1 penny)

●○○○○ Ask about the number of dimes and pennies for amounts under $1.00. Example: *How many dimes and pennies in 83¢?* (8 dimes and 3 pennies)

●●○○○ Ask about the number of dimes and pennies for amounts over $1.00. Example: *How many dimes and pennies in $2.47?* (24 dimes and 7 pennies)

●●●○○ Ask about the number of nickels and pennies for amounts under and over $1.00. Examples: *How many nickels and pennies in 56¢?* (11 nickels and 1 penny) *How many nickels and pennies in $1.13?* (22 nickels and 3 pennies)

●●●●○ Ask about the number of quarters and pennies for amounts under and over $1.00. Examples: *How many quarters and pennies in 56¢?* (2 quarters and 6 pennies) *How many quarters and pennies in $1.45?* (5 quarters and 20 pennies)

 CCSS 2.MD.8   115

Again, How Many?

Ask: *If I have **76¢**, what coins do I have if I have . . .*

1. *only dimes and pennies?* (7 dimes, 6 pennies)

2. *only nickels and pennies?* (15 nickels, 1 penny)

3. *quarters, dimes, and pennies?* (1 quarter, 5 dimes, 1 penny; 2 quarters, 2 dimes, and 6 pennies; and so on)

4. *quarters, dimes, nickels, and pennies?* (2 quarters, 2 dimes, 1 nickel, 1 penny; 1 quarter, 4 dimes, 2 nickels and 1 penny; and so on)

●○○○○ Choose amounts under 50¢.

●●○○○ Choose amounts under $1.00.

●●●○○ Choose amounts over $1.00.

Telling Time

Ask: *About what time is it?*

⬤〇〇〇〇 Have children respond to the nearest hour or half hour.

⬤⬤〇〇〇 Have children respond to the nearest 5 minutes.

⬤⬤⬤〇〇 Have children respond to the nearest minute.

⬤⬤⬤⬤〇 Have children respond to the nearest minute. Ask how long it is in minutes until the next hour.

Measures of Time

Brainstorm with children about activities they can do in . . .

●○○○○ about a minute. (Getting a drink of water, getting dressed)
about 30 minutes. (Watching a television show)
about 1 hour. (Math class)

●●○○○ about 15 minutes. (Recess, playing a few rounds of a math game)
about 45 minutes. (Art class, lunch period)
about 2 hours. (Watching a movie, playing an entire
basketball game)

Try timing classroom routines and activities throughout the day and add to
the lists of activities given above.

Washing and Drying Clothes Number Stories

It takes my washer about 30 minutes to wash clothes, and it takes my dryer about 45 minutes to dry clothes.

● ○ ○ ○ ○ About how much longer does it take me to dry clothes than to wash them? (15 min)

● ● ○ ○ ○ About how long will it take me to wash and dry a load of clothes? (75 min, or 1 hr 15 min) To wash and dry 2 loads of clothes? (150 min, or 2 hr 30 min)

● ● ● ○ ○ My neighbor's washing machine washes clothes about 5 minutes faster than mine, but he has the same dryer as I do. How long will it take him to wash and dry a load of clothes? (70 min, or 1 hr 10 min)

● ● ● ● ○ I have about 2 hours of free time. How many loads of laundry can I wash and completely dry? (Only 1 load of laundry; I would not have time to dry the second load.) If I have about 3 hours of free time? (2 loads)

CCSS 1.OA.1, 1.NBT.4, 2.OA.1, 2.MD.7, 3.NBT.2, 3.MD.1

Game Time Number Stories

Samantha and Aisha played a game for about 45 minutes.

○○○○ For about how long did each of them play? (45 min) Did the girls play for more or less than 1 hour? (Less) How much less than an hour? (15 min less)

●●○○○ For what fraction, or portion, of an hour did the girls play? $\left(\frac{45}{60} \text{ or } \frac{3}{4} \text{ of an hour}\right)$

●●●○○ If they wanted to play the same game twice, about how long would it take them? $\left(90 \text{ min or } 1\frac{1}{2} \text{ hr}\right)$ If they wanted to play the same game 3 times, about how long would it take them? $\left(135 \text{ min or } 2\frac{1}{4} \text{ hr}\right)$

Time Zones*

Say: *The time in New York City is one hour later than the time in Chicago. The time in Los Angeles is two hours earlier than the time in Chicago. If it is* **3:00 P.M. in Chicago,** *what time is it in* **New York City?** (4:00 P.M.)

●○○○○ Tell children a time in Chicago; ask them to give the time in New York City or Los Angeles.

●●○○○ Tell children a time in New York City; ask them to give the time in Los Angeles and the time in Chicago. Also tell children a time in Los Angeles; ask them to give the time in New York City or Chicago.

*Use time comparisons from your time zone.

Elapsed Times

Say: *It is about **2:30 P.M.***

●○○○○ Ask: *What time will it be in an hour?* (3:30 P.M.) *What time will it be in a half hour?* (3:00 P.M.) *What time will it be in two hours?* (4:30 P.M.)

●●○○○ Ask: *What time was it one hour ago?* (1:30 P.M.) *What time was it a half hour ago?* (2:00 P.M.)

●●●○○ Ask: *What time will it be in 15 minutes?* (2:45 P.M.) *What time will it be in 45 minutes?* (3:15 P.M.)

●●●●○ Ask: *What time will it be in 12 minutes?* (2:42 P.M.) *What time was it 6 minutes ago?* (2:24 P.M.)

Repeat with other start times.

Equivalent Measures

Say: *After I tell you an amount, find the equivalent amount.* (Each of the following variations can be made more or less difficult by altering the size of the numbers and how easily they divide into desired unit amounts.)

Variation 1 Ask about money equivalents: *4 nickels equals how many dimes?* (2 dimes) *2 dollars equals how many quarters?* (8 quarters)

Variation 2 Ask about time equivalents: *1 week is how many days?* (7 days) *1 day equals how many hours?* (24 hr) *1 month is about how many days?* (28, 29, 30, or 31 days) *2 days equals how many hours?* (48 hr) *9 minutes equals how many seconds?* (540 sec)

Variation 3 Ask about distance equivalents: *12 feet is the same as how many yards?* (4 yd) *100 centimeters is the same as how many meters?* (1 m)

Mass Number Stories `Number Stories`

●○○○○ A paper clip has a mass of 1 gram. What is the mass of 10 paper clips? (10 g) 1,000 paper clips? (1,000 g or 1 kg)

●●○○○ 1,000 grams are in 1 kilogram. What is the mass of 1 kilogram of apples in grams? (1,000 g) 2 kilograms of apples? (2,000 g)

●●●○○ Libby has half a kilogram of apples, and she buys another half kilogram. How much mass do her apples have all together? (1,000 g or 1 kg)

Short and Tall Women* | Number Stories

The shortest woman on record, Pauline Musters, was about 1 foot 11 inches tall. The tallest woman, Zeng Jinlian, was about 8 feet 2 inches tall.

●○○○○ To the nearest foot, how tall was Pauline Musters? (2 ft) To the nearest foot, how tall was Zeng Jinlian? (8 ft)

●●○○○ If Zeng and Pauline stood back to back, what would be the vertical distance between the tops of their heads? (About 6 ft)

●●●○○ The average height for American women is about 5 feet 4 inches. How much taller than the average woman was Zeng? (2 ft 10 in.) How much shorter than the average woman was Pauline? (3 ft 5 in.)

*You may want to display these heights.

Measurement and Data

Bargain Shopping Number Stories

Which is the bargain?

Pretzels: 10¢ each **or** 3 for a quarter (3 for 25¢)

Crackers: a nickel each **or** 5 for a quarter (Neither. Both cost the same.)

Orange juice: a small carton for 50¢ **or** a large carton for 75¢ (Insufficient data to answer the question. You don't know the amounts of juice in "small" and "large" cartons.)

Strawberries: 15¢ each **or** 6 for $1.00? (15¢ each)

Pizza: $1.50 per slice **or** a 6-slice pizza for $8.50? (6-slice pizza)

Shapes of Signs

Say: *Imagine you are in a car or a bus. What are the shapes of some road signs that you might see?*

Children may respond with the following shapes:

triangle (yield sign)

rectangle (speed limits, warning signs, street names, and so on)

octagon (stop sign)

circle (railroad crossings)

Ask: *Why do you think road signs are different shapes?*

Imagining Shapes

Say: *Imagine a baseball game. What are some shapes you might see?*

Variations Tell children to imagine an event, an activity, a place, or a thing with which they might be familiar, such as a football game, a hockey game, a basketball game, the zoo, the circus, the playground, a bedroom, a kitchen, the grocery store, a lunch, and so on. Ask: *What are some shapes you might see?*

Identifying Line Segments

Say: *Look around the **room** and tell me where you see . . .*

●○○○○ *line segments.* (Edge of a book, the window, the second hand on the clock, a shelf, corners where walls meet or where walls meet the ceiling, and so on)

●●○○○ *line segments that are parallel.* (Sides of a bookcase, sides of a door, and so on)

Making Shapes and Angles

●○○○○ Say: *Use your fingers, arms, or body to form a **circle**.* (Also tell children to form other shapes such as triangles, rectangles, and so on.)

●●○○○ Say: *With a group of three or four other children, form a **circle** either standing up or lying down.* (Also tell children to form polygons such as triangles, rectangles, pentagons, hexagons, septagons, and octagons.)

●●●○○ Say: *Using your arms or fingers, form a right angle. Now form parallel line segments.*

Variation Ask another group of three or four children to form a different shape surrounding the original group shape.

Walking Shapes

Say: *I will choose one child to walk on an imaginary outline of a flat shape on the floor while the rest of us guess the shape.*

●○○○○ Have the chosen child select any shape to "walk."

●●○○○ Have the chosen child follow the instructions of another child, who will describe how to walk the shape. Example: To describe a triangle, the child might say, "Walk four paces, turn to the right, walk four more paces, turn to the right, and walk back to the starting point."

●●●○○ Have the chosen child follow the instructions of another child, who will describe how to walk the shape using instructions that include directions, degrees, and paces. (Display north, south, east, and west labels in your classroom.) Example: To describe a square, the child might say, "Begin facing north, walk three paces, turn toward the east at a 90° angle, walk three more paces, turn right at a 90° angle to the south, walk three more paces, turn right at a 90° angle toward the west, and walk three more paces."

Describing Shapes

Say: *I'm thinking of a **2-dimensional** shape that has **3 sides** and **3 angles**. Does anyone know what the shape is?* (A triangle)

●○○○○ Think of and describe a 2-dimensional shape: circle, triangle, square, and so on. Ask children to identify it.

●●○○○ Think of and describe a 3-dimensional shape: sphere, pyramid, cylinder, cube, prism, or cone. Ask children to identify it. Example: *I'm thinking of a 3-dimensional shape that looks like a soup can. What is it?* (Cylinder)

Class Shapes and Line Segments

Say: *Form the shape of a **square** with everyone in the class.*

●○○○○ Tell the class to form a 2-dimensional shape: circle, square, triangle, and so on.

●●○○○ Tell the class to form two parallel line segments; two perpendicular line segments; two intersecting, non-perpendicular line segments; or a right angle.

Notes

List of Activities by Page

Basic Routines 1
    Numbers Before and After 3
    Numbers Between 4
    Counts and Skip Counts 5
    Ordinal Numbers 6
    Count by 10s and 100s 7
    What Do I Do? 8
    Using Combinations of 10 9
    Arithmetic Facts 10
    Name Collections (Equivalents) 11
    More Name Collections (Equivalents) 12
    Multistep Problems 13
    How Many 10s, 100s, 1,000s? 14
    "What's My Rule?" 15
    Number Stories 16
    Shapes Around Us 17
    Geometry "I Spy" 18
    How Many Cents? 19
    Place Value 20
    Finding 10 More and 10 Less 21

Minute Math+ Topics 23

Operations and Algebraic Thinking 25
    Missing Parts in Sums and Differences.................................... 25
    Sleep Needs 26
    Thunder.. 27
    Rug Measures 28
    Old Milk... 29
    Dreams 30
    Classroom Counts 31
    Hot Dog Buns... 32
    Baby Penguin Meals 33
    Fact Families.. 34
    Making Omelets 35
    Bags of Apples 36
    Toilet Flushes.. 37
    A Snail's Pace 38
    Making Apple Juice 39
    Baby Weights... 40
    Cooking-Oil Consumption 41
    Planting Flower Bulbs 42
    Digit Arithmetic.. 43
    Elephant Sleep.. 44

Hibernation 45
Making Muffins 46
Types of Bears 47
Pumping Blood 48
Basketball Player Heights 49
Walking Rates 50
Paper in Trash Heaps 51
Secret Numbers 52
Penguins vs. Human Swimmers 53
Animal Litters 54
Animal Weights 55
Growing Eyelashes 56
Milk Consumption 57
Watching Sheep 58
Granola Bar Sale 59
Dinosaur Sizes 60
Double, Triple, Quadruple 61
Age: People vs. Mice 62
A Dog's Age 63
Another Dog's Age 64
Whale Speeds 65
Baseball Speeds 66

Number and Operations in Base Ten 67

Numbers After and Before 67

Whole Numbers Between 68

The Oldest Living Animal 69

Outdoor Temperatures 70

Missing Numbers 71

How Many? 72

Elephant Meals 73

New Books 74

Continue the Sequence 75

Repeated Digits 76

Tall Buildings 77

Breathing 78

Numbers with *n* Digits 79

Easier Numbers 80

Bird Egg Sizes 81

Hummingbird Wing Flaps 82

Creating Numbers 83

Guess My Number 84

Extreme Temperatures 85

Alligator Eggs 86
Adding and Subtracting Multiples of 10 87
Multiplication Properties of 10. 88
Airplane Speed vs. Human Speed. 89
Bacteria Growth 90
Siblings 91
Estimating Differences. 92
Popping Corn 93
Earth's Orbit 94

Number and Operations — Fractions **95**
Parts in a Whole 95
Body Muscle 96
Making Orange Juice 97
Parts 98
Body Water Weight. 99
Making Pancakes 100
Measurement Fractions 101
Koala Birth Weights 102
A Pint Is a Pound. 103
Water in an Elephant Trunk 104

Measurement and Data. **105**

   Money and Measure Counts. 105

   Measuring Tools. 106

   Informal Measuring Tools 107

   Making a Dollar. 108

   Money Exchanges 109

   More Money Exchanges 110

   The Length of a Dollar Bill 111

   Koala Bears 112

   Coin Equivalents. 113

   Money and Time 114

   How Many?. 115

   Again, How Many?. 116

   Telling Time. 117

   Measures of Time. 118

   Washing and Drying Clothes 119

   Game Time. 120

   Time Zones 121

   Elapsed Times. 122

   Equivalent Measures 123

   Mass Number Stories 124

   Short and Tall Women. 125

   Bargain Shopping. 126

Geometry **127**
   Shapes of Signs.. 127
   Imagining Shapes... 128
   Identifying Line Segments 129
   Making Shapes and Angles 130
   Walking Shapes.. 131
   Describing Shapes.. 132
   Class Shapes and Line Segments............................... 133

Key to Sources

Minute Math+ page	Source
27	Sandow, p. 18
30	Smith and Moore, p. 84
37	Smith and Moore, p. 74
40	Parker 1987, p. 23
42	Parker 1983, p. 20
44	Balfanz, p. 67
45	McFarlan, p. 36
45	Stephen, p. 152
47	Great Bear Foundation website
48	Parker 1987, p. 41
49	McFarlan, p. 344
50	Smith and Moore, p. 34

Minute Math+ page	Source
51	United States Environmental Protection Agency website
54	Burton, p. 84
55	Diagram Group, p. 119
56	Smith and Moore, p. 10
60	Diagram Group, p. 65
63	Parker 1987, p. 282
64	Parker 1987, p. 282
65	Stephen, p. 306
66	Smith and Moore, p. 199
69	National Geographic website
73	Smith and Moore, p. 99
74	Bowker website

Minute Math+ page	Source
74	American Library Association website
77	John Hancock Center Observatory
78	Parker 1987, p. 48
81	Diagram Group, p. 69
82	Sandow, p. 12
85	Guinness World Records website
86	Smith and Moore, p. 107
93	Parker 1987, p. 15
96	Parker 1987, p. 9
97	Parker 1983, p. 9
104	Smith and Moore, p. 101
125	Guinness World Records website

Bibliography

American Library Association. "Number of Children's Books Published." Accessed October 9, 2014. http://www.ala.org/tools/number-childrens-books-published.

Balfanz, Robert. *Do Elephants Eat Too Much?* Evanston, IL: Everyday Learning Corporation, 1992.

Bowker. "Output Report for 2002–2013." Accessed October 9, 2014. http://bowker.com.

Burton, Maurice, ed. *The World Encyclopedia of Animals.* New York: Funk & Wagnalls, 1972.

Diagram Group. *Comparisons.* New York: St. Martin's, 1980.

Great Bear Foundation. "The Eight Bear Species of the World." Accessed October 9, 2014. http://greatbear.org/bear-species/.

Guinness World Records. "Highest Recorded Temperature." Accessed October 9, 2014. http://www.guinnessworldrecords.com/world-records/3000/highest-recorded-temperature.

———. "Lowest Temperature – Inhabited." Accessed October 9, 2014. http://www.guinnessworldrecords.com/world-records/3000/lowest-temperature-inhabited.

———. "Shortest Woman Ever." Accessed October 9, 2014. http://www.guinnessworldrecords.com/world-records/1/shortest-woman-ever.

———. "Tallest Woman." Accessed October 9, 2014. http://www.guinnessworldrecords.com/world-records/3000/tallest-woman.

John Hancock Center Observatory. Information provided to visitors. Chicago.

McFarlan, David A., ed. *Guinness Book of World Records 1990.* New York: Sterling Publications, 1989.

National Geographic. "Six of the World's Longest-Lived Animals." Accessed October 9, 2014. http://newswatch.nationalgeographic.com/2014/02/04/6-of-the-worlds-longest-lived-animals/.

Parker, Tom. *Rules of Thumb*. Boston: Houghton Mifflin, 1983.

———. *Rules of Thumb 2*. Boston: Houghton Mifflin, 1987.

Sandow, Stuart A., ed. *Durations: The Encyclopedia of How Long Things Take*. New York: Avon, 1977.

Smith, Richard, and Linda Moore, eds. *The Average Book*. New York: The Rutledge Press, 1981.

Stephen, David, ed. *Encyclopedia of Animals*. New York: St. Martin's, 1968.

United States Environmental Protection Agency. "Municipal Solid Waste Generation, Recycling, and Disposal in the United States: Facts and Figures for 2012." Accessed October 9, 2014. http://www.epa.gov.